This book is to be returned on or before

St Clements
class 2

TEAMWORK
Fire Service

Philippa Perry and Stephen Gibbs

Wayland

Imprint details are on page 31.

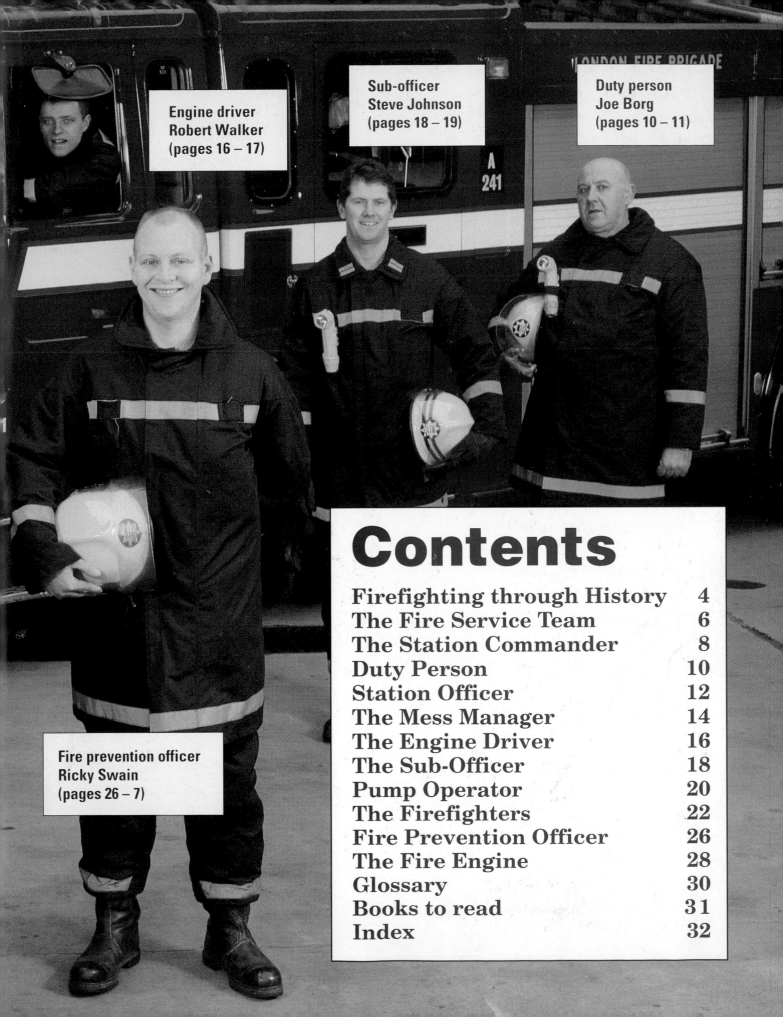

Engine driver
Robert Walker
(pages 16 – 17)

Sub-officer
Steve Johnson
(pages 18 – 19)

Duty person
Joe Borg
(pages 10 – 11)

Fire prevention officer
Ricky Swain
(pages 26 – 7)

Contents

firefighting through History

Roman firefighters

One of the earliest firefighting teams was set up by Emperor Augustus in Rome, in about 27 BC. These firefighters were called *vigiles* and they patrolled the streets checking for fires and putting out them out.

Fire hazard

During the Middle Ages, most buildings were made of wood, which caught alight very easily, so fires were common. In 1189, King Richard I passed a law which stated that all houses had to have a barrel of water outside at all times to help to put out any fires.

London on fire

In 1666, the Great Fire of London destroyed 13,000 houses and eighty-seven churches. Insurance companies had to pay for some of the buildings that burnt down in the fire. So, after this disaster they organized private fire brigades to protect the buildings they insured.

The Fire Brigade

The modern fire service started when the Metropolitan Fire Brigade was set up in 1866, in London. It used steam-driven water pumps on carts pulled by horses. The steam engine worked machinery that pumped up the water. Using the force of the pumped water was better for putting out fires than pouring on buckets and barrels of water.

This picture shows a team from the Metropolitan Fire Brigade racing to the scene of a fire. ▶

The first petrol- or diesel-powered engines appeared in the early 1900s. Since then, many improvements have been made in the equipment and methods used to fight fires. ▶

Fact-file

- The Latin word *vigile* means 'watch'.
- Incredibly, only six people died in the Great Fire of London in 1666.

The Fire Service Team

Every year, fire stations all over Britain answer more than 1,000,000 emergency calls. It may be a tower-block on fire, a smoking dustbin or simply a cat stuck up a tree. Whatever the call, the fire service has to be ready for action twenty-four hours a day, 365 days of the year.

Every day, firefighters risk their lives to save people and protect property from fires. They have one of the most dangerous jobs in the community. Have you ever wondered what it takes to be a firefighter? It takes a lot of hard work and years of experience, with everyone at the fire station having a particular part to play. When a fire is blazing out of control, good teamwork helps to save lives.

Every second counts

When you dial the emergency number '999' to report a fire, your call goes automatically to the central fire department for your area. An operator types the exact location and any other details about the fire into a computer. The computer then sends the information by fax to the nearest fire station in the area.

The Soho fire station, London

Soho is an area of central London. It is one of the most crowded areas of the capital – as a result, its fire station is one of the busiest in the country.

◀ **Two engines pulling out of Soho fire station on an emergency call-out.**

These are some of the members of 'White Watch' and 'Red Watch' at Soho fire station. Find out more about each person's job and how they work as a team on the relevant pages of this book. ▶

▲The station officer is in charge of training the firefighters in his watch. Here he is supervising a training drill in the exercise yard.

The station officer needs to be a good listener, because sometimes the members of the watch come to him with their problems. Being a firefighter can be a very stressful job, especially when there is a very big fire with lots of people hurt.

Fact-file

● Soho station's firefighters get called out about 3,000 times a year, and sometimes as many as thirty times in twenty-four hours.
● The station officer organizes regular training courses and classes on all the most up-to-date firefighting skills and equipment.

The Mess Manager

The mess manager is in charge of all the meals and makes sure the firefighters are properly fed during their watch. Fighting fires is a very physical, energetic job, so good food is very important!

Every morning, the mess manager buys all the food for the day. The members of the watch eat at least five meals a day – breakfast, lunch, dinner, plus a mid-morning and an afternoon snack.

The mess manager is a trained firefighter like all the others in the watch. When the emergency bell sounds, there is no time to lose. All the firefighters – including the mess manager – get going immediately, even if they are in the middle of a meal!

The mess manager, Eddie Martin, is in charge of the kitchens and the meals cooked for his watch. ▶

Eddie Martin
'Before I started this job I didn't think I could cook at all, but now a meal for eighteen hungry mouths five times a day is no problem.'

Fact-file
- Each watch munches its way through about ten sacks of potatoes, fourteen loaves of bread, three and a half kilogrammes of meat and nine litres of milk every week.
- The word mess comes from the French word *mets* which means 'dish'.

▲ **Hungry firefighters waiting for a hot meal.**

The mess manager is a trained firefighter just like everyone else in the watch. ▶

The Engine Driver

The engine driver's job is to get the members of the watch to the emergency as quickly and safely as possible. He or she has to know all the roads in the area very well.

Once at the scene, the driver can help the other members of the team to tackle the blaze. Usually the driver is also the pump operator in charge of the water pumps on the engine (see pages 20 – 21).

The engine driver, Robert Walker, drives quickly and safely to the scene of a fire or accident. ▼

When a call slip comes through, the duty person gives the driver exact directions to the fire. Some streets are too narrow for fire engines – so the driver has to know alternative routes. The driver switches on the engine's fire-bell to warn other vehicles and people to clear the roads.

Twice a day the driver checks and cleans the fire engine thoroughly from top to bottom, making sure the tyres are full of air and there is enough fuel in the tank.

▲ **The driver checks over the equipment on the engine with the sub-officer, Steve Johnson (see pages 18 – 19).**

Fact-file

● In a built-up area like Soho the top speed the engine reaches is about 80 k.p.h. The fastest fire engine in the world is the Jaguar XJ12 'Chubb firefighter'. It reaches speeds of over 209 k.p.h.
● When drivers of other vehicles hear the fire-bell of an engine they must pull over to allow the engine to get through.
● All fire-engine drivers have to pass a test for a special licence to drive large vehicles, called a 'Heavy Goods Vehicle (HGV) licence'.

Robert Walker
❝ I don't drive at top speeds because that would be too dangerous, especially in somewhere like central London. But I try to get to the emergency as quickly as possible. ❞

The driver at the wheel. ▶

The Sub-Officer

Once the team has put a fire out, it is the sub-officer's job to try and find out how it started. He or she does this by interviewing witnesses and examining the scene of the accident very carefully for clues. Back at the fire station the sub-officer fills in a report with all the details about the fire, such as where it happened and how it may have started.

The sub-officer is the second-in-command for the watch and takes orders from the station officer. It is often his or her job to train all the firefighters. Firefighters have to do lots of different training exercises, ranging from written tests to learning how to use ladders on very tall buildings.

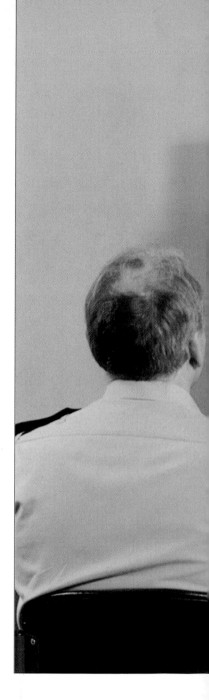

▼ **The sub-officer, Steve Johnson, writes up a report after each fire or accident.**

Training other firefighters is an important part of the sub-officer's job. ▶

The sub-officer takes orders from the watch's station officer and passes on instructions to the rest of the watch members. ▶

Steve Johnson
❝ To be good at this job, you've got to be able to make a decision and persuade everyone else it's a good idea! ❞

Blackboard text:

FIREFIGHTER C.A.SET

1·7 kg   STAINLESS ST... Y PLATE
2250 litres @ 210 bar – MINI... ...40 bar
PRESSURE – APPROX 46 mins ...tion
...RIC SEAL o... ...MASK – APP...
...E REDUCERS – WARNING...
...NE – 6M – ADSU – 2 CELL...
...RIGADE – STN – WEARERS NAM...
...PRESSURE – ... – DOSIMETER
...TYPE

NO S...

Fact-file

• A fire station is run very much like an army barracks. Training exercises are called 'drills' and watch leaders are 'officers'.

• To control fires quickly firefighters need to be trained to take orders and follow the instructions properly.

19

Pump Operator

Massive amounts of water are needed to put out a fire. The pump operator makes sure that the firefighters do not run out of water.

1. The fire engine can carry only a limited amount of water. But it can pump a lot of water from the mains pipes which run under the street and supply water to local buildings and houses.

2. As soon as the fire engine arrives at the scene of the blaze, the pump operator finds the nearest water point (called a hydrant) on the street. He or she then fixes a hose from the fire engine to this hydrant.

3. The firefighters can now use the hose to put out the fire. If they need more pressure, they radio down to the pump operator.

4. Petrol fires can be put out with foam. Fire engines have a special tank which contains a liquid that, when mixed with water, produces a white foam.

▲ **Chris Andy is the person who operates the water pumps on the engine.**

▼ **Firefighters getting a fire hose under control.**

Hose out of control!

When the pressure is high, the water shoots out of the hose at an incredible speed. Sometimes hoses can get out of control. It can take as many as three firefighters to bring the hose back under control. They do this by crawling along the length of the pipe on their hands and knees and then lying on the end!

Chris Andy

'The driver usually operates the pump but all the members of the watch learn all the jobs, including that of the pump operator.'

Chris Andy keeps in contact with the other firefighters by 'walkie-talkie'. They tell him how much water pressure is needed and then he adjusts the controls. ▼

Fact-file

- The water tank on the engine holds 1,365 litres of water.
- An average hose is 25-metres long.
- The fire engine has suction hoses which can pump water out of lakes, rivers and even swimming pools.
- Just like on many garden hoses, there are controls so that the water can come out in either a fine spray or a jet.
- The fire engine with the greatest pumping power is the Oshkosh firetruck used for aircraft and runway fires at airports. It pumps out 189,000 litres of foam from two pipes in just two and a half minutes.

The Firefighters

All the members of the watch are trained firefighters. It is their job to put out fires and save lives and property. Firefighters always work as a team. In an emergency, they know that they can rely on any other member of the team.

At the fire station
Firefighters have to be ready to rush to the scene of a fire at any time, but when they are not fighting fires they do not just sit around twiddling their thumbs. Much of the shift is spent doing training exercises in the yard with the sub-officer and station officer.

The rest of their time is spent cleaning and checking the engines and equipment. Also, boring jobs like sweeping the fire station floor have to be done.

Saved by the bell
When the alarm bell sounds the firefighters leave what they are doing and rush to their engines. While the fire engines speed through the streets, the firefighters finish putting on their protective splash suits (see page 24) and check their breathing apparatus (see above right). Just like divers, firefighters wear special oxygen tanks to help them to breathe. Each tank holds enough air to last for forty-six minutes, so when the firefighters are working in a blazing building they must regularly check how much air is left in their tanks.

At night the training room is used as a dormitory for the firefighters on night-watch so that they can sleep between call-outs.

Firefighter Nigel February checking his breathing apparatus.

When they arrive

The firefighters' first job is always to stop the fire from spreading. Often this involves spraying nearby buildings with water to stop them from catching fire too.

If there are people stranded high up in a building, the firefighters use a thirty-metre-high aerial ladder platform to rescue them. The platform can be raised or lowered by controls on its engine.

Firefighters always work in pairs. They move through the building breaking down doors if necessary and spraying water on to the blaze. Sometimes the inside of the building is so black with smoke that they cannot see at all. But this does not stop them. They just have to feel their way around the walls. They keep in contact all the time with the station officer and the other firefighters by radios, called 'walkie-talkies'.

To get to the fire engines quickly the firefighters slide down poles instead of using stairs when a call-out is received.

23

Top gear

Fires are very dangerous and the firefighters need protection, not only from the heat and falling debris, but also from dangerous substances, such as burning chemicals. So, depending on the type of fire they are dealing with firefighters have to wear special clothes for the task.

Splash suits: for most fires, firefighters wear splash suits. These suits are made of black and yellow heat-resistant plastic. They have a long jacket and high-waisted trousers, with braces to help them to stay up. The suits keep the firefighters dry and protect them from the heat of a fire. The firefighters also wear heavy boots, gloves and yellow hard hats.

Chemical protection suits: these are worn when the firefighters are dealing with any fire where there might be dangerous chemicals around. In these suits, every part of the firefighters' bodies is protected from poisonous substances.

Heat-reflective suits: these suits are coated with aluminium to reflect heat. They are completely fire-resistant, which means that when wearing one a firefighter can walk through fire and not get burnt. They are most often used for aircraft fires.

Carmen Curran
'I feel really excited when I reach a fire. Suddenly, every action counts and all the training we've done as a team takes over.'

The firefighters' splash suits, boots and helmets for each watch are stored neatly and tidily in the locker-room. ▶

▲ Two other firefighters are helping Carmen Curran to put on a chemical protection suit. The suit fits over her usual firefighting gear of splash suit, hard hat and breathing apparatus.

Fact-file

- Firefighters have to be at least eighteen years old.
- The minimum height is usually 1.67 metres and the maximum height is 1.93 metres.
- Firefighters have to be physically fit and have good eyesight.
- New firefighters are sent on a twelve-week training course at a special college before they can join a fire station.

The best way to stop a fire is to prevent it from happening in the first place. The fire prevention officer inspects local public buildings, such as schools, hotels, restaurants, cinemas and offices. He or she teaches people about fire safety and what to do if they discover a fire in their building.

In Britain, there is a special fire safety code for all public buildings. This says that they must have a certain number of fire extinguishers and fire exits. The fire prevention officer checks that these rules are being obeyed.

▲ The fire prevention officer, Ricky Swain, advises people about fire safety.

Rules in the case of fire

1. Leave the building immediately.

2. Never open a door that feels hot. Before opening any door, put your hand on it. If it feels hot, the fire on the other side may be blazing fiercely.

3. Crawl on the floor when going through a smoke-filled area. Smoke rises, so it will be thinnest nearer the floor.

4. Do not run if your clothes catch fire. Running spreads the flames. Roll on the floor and smother the flames.

5. Do not return to the building for any reason. After you have got out, call the fire service on the '999' emergency number.

Ricky Swain
‘ Many fires could be avoided – my job is to help people to be aware of the things that could be a fire hazard. ’

◀ Giving advice about fire hazards is an important part of the fire prevention officer's job.

▲ Ricky Swain checking that a fire extinguisher is in working order.

Fact-file

- Most new office buildings have sprinkler systems. If a fire starts, the heat and smoke set off the sprinklers immediately, showering the room with water.
- The main causes of fire in the home are faulty electrical plugs, cigarettes left burning, chip pans catching alight and open fires left without a guard or with material too close to them.

The Fire engine

The fire engine has a thirteen-metre and a nine-metre long extension ladder on top.

The hoses, some as long as twenty-five metres, are stored neatly in coils.

A fire extinguisher is kept on the engine.

Spare chemical protection suits and breathing apparatus are stored here.

The fire-bell and lights. When the engine is on a call-out the driver sounds the bell and switches on the flashing lights to warn other road users.

The 'cab' at the front of the engine, where the driver and the sub-officer sit. The other firefighters sit in a separate section behind the cab.

Glossary

back-up Other firefighting teams or emergency services, such as the ambulance or police, who go along to the scene of an accident to help the team already there.

breathing apparatus The equipment which firefighters wear so that they can breathe and work safely in smoke-filled areas. A special mask fits over the face to stop the smoke reaching the firefighter's eyes, nose or mouth. The firefighter breathes oxygen through a mouth-piece and tube connected to an oxygen tank which is carried on his or her back.

brigade A group of people organized for a certain task.

call-out When firefighters are sent to the scene of an accident or fire.

debris The material, such as bricks, wood and metal, from a building which has been destroyed.

emergency A sudden event, such as an accident or fire, which requires urgent action.

extinguishers Hand-operated devices, usually filled with foam, which can be used to put out small fires.

false alarm When the result of a call-out is not serious.

fax A written message sent or received through a facsimile 'fax' machine. A fax machine at one end of a telephone line changes the writing on the original message into a series of signals. These signals are sent down the telephone line and received by another fax machine which produces the message in writing again.

heat-resistant A substance or material which stops the heat of a fire from reaching the body.

insurance To insure (or guard) against loss of property by fire, accident or theft by making payments to a person or company who promises to pay out an agreed amount of money to replace the lost property.

Middle Ages The period between the fifth and fifteenth centuries AD.

poisonous Describing a substance which is dangerous if it is taken into the body.

pressure A force, such as air pressure, pressing on a surface. When a controlled amount of air pressure is introduced into the water or foam storage tanks on a fire engine the liquid is forced out at great speed.

prevent To stop something from happening.

protective Something giving safety or cover.

steam engine An engine which burns a material, such as coal, to heat water so that steam is produced to power machinery.

suction To draw up a substance, such as water, by sucking.

Books to read

Fire Engines by Norman Barrett (Franklin Watts, 1991)

Fire Engines by R.J. Stephen (Franklin Watts, 1991)

Images of Fire by Neil Wallington (David and Charles, 1989)

Firefighter (Living Dangerously series) by Neil Wallington (Wayland, 1992)

Titles in the series
Building Site
Fire Service
Hospital
Newspapers
Police Service
Post Office

Series Editor: Geraldine Purcell
Series Designer: Loraine Hayes

© Copyright 1994 Wayland (Publishers) Limited

First published in 1994 by Wayland (Publishers) Limited
61 Western Road, Hove, East Sussex BN3 1JD, England.

British Library Cataloguing in Publication Data

Perry, Philippa
Fire Service. – (Teamwork Series)
I, Title II. Gibbs, Stephen III. Series
363 . 3780941

ISBN 0 7502 1101 6

DTP Design by Loraine Hayes
Printed and bound in Italy by Rotolito Lombarda S.p.A

Useful Addresses

Home Office Fire Department
Horseferry House
Dean Ryle Street
London SW1P 2AW
Tel: 071 582 3811

London Fire Brigade Museum
Winchester House
92 Southwark Bridge Road
Southwark,
London SE1 0EG
Tel: 071 587 2894

Index

Acknowledgements
The authors and publisher would like to thank all the firefighters at Soho fire station, London, for their co-operation in the making of this book. Special thanks goes to the members of 'White Watch' and 'Red Watch'.

Picture acknowledgements
All the photographs in this book were taken by Andrew Perris and Paul Mattock, APM Studios, except for the following: Mary Evans Picture Library 4 – 5 (all photographs).